Animals in Savannas

Julie Murray

Abdo Kids Junior
is an Imprint of Abdo Kids
abdobooks.com

abdobooks.com

Published by Abdo Kids, a division of ABDO, P.O. Box 398166, Minneapolis, Minnesota 55439.

Printed in China

052020

092020

Photo Credits: iStock, Shutterstock

Production Contributors: Teddy Borth, Jennie Forsberg, Grace Hansen

Design Contributors: Candice Keimig, Pakou Moua, Dorothy Toth

Library of Congress Control Number: 2019955591

Publisher's Cataloging-in-Publication Data

Names: Murray, Julie, author.

Title: Animals in savannas / by Julie Murray

Description: Minneapolis, Minnesota : Abdo Kids, 2021 | Series: Animal habitats | Includes online resources and index.

Identifiers: ISBN 9781098202101 (lib. bdg.) | ISBN 9781098203085 (ebook) | ISBN 9781098203573 (Read-to-Me ebook)

Subjects: LCSH: Animals--Habitations--Juvenile literature. | Habitat (Ecology)--Juvenile literature. | Savanna animals--Juvenile literature. | Savannas--Juvenile literature. | Savanna ecology--Juvenile literature.

Classification: DDC 591.52--dc23

Table of Contents

Animals in Savannas 4

More Animals in Savannas. 22

Glossary. 23

Index 24

Abdo Kids Code. 24

Animals in Savannas

Many animals live in a savanna. It is hot and **dry**.

Zebras live in a **herd**.

This keeps them safe.

Giraffes are tall. They can stand 20 feet (6.1 m) high!

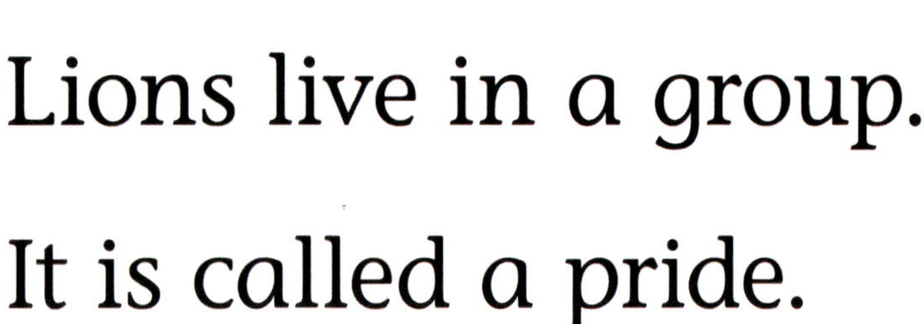

Lions live in a group.

It is called a pride.

Cheetahs are fast. They can run up to 75 mph (120 km/h)!

Impalas have horns. The horns are long and curvy.

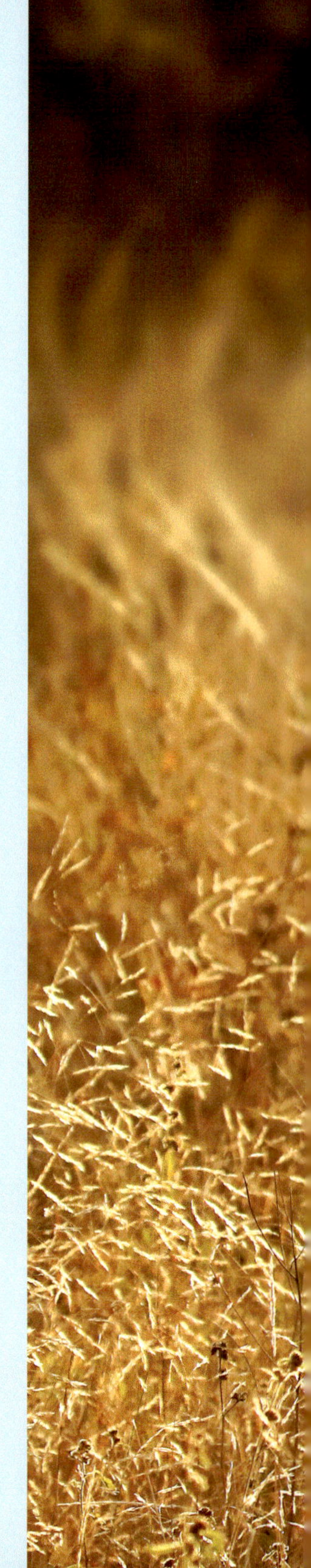

Ostriches are big birds.

They cannot fly.

Rhinos lay in the mud.

This cools them off.

Elephants have big trunks.

They use them to drink water!

More Animals in Savannas

black mamba

gazelle

meerkat

wildebeest

Glossary

dry

having little or no rain.

herd

any group of wild animals that feed and travel together.

Index

cheetahs 12

climate 4

elephants 20

giraffes 8

impalas 14

lions 10

ostriches 16

rhinoceroses 18

safety 6

zebras 6

Visit **abdokids.com** to access crafts, games, videos, and more!

Use Abdo Kids code

AAK2101

or scan this QR code!